Weather Instruments and How to Make Them

By: Gerard W. Telmosse

ISBN: 978-1-105-78511-5

Interior Book Design and Layout by
www.angelediting.com

Cover Design by
Gerard W. Telmosse

Published through Lulu Press, Inc.

DEDICATION

I would like to dedicate "Weather Instruments and How to Make Them" to Dr. Mitch Batoff. Mitch has been both an inspiration and a mentor over the past twenty years. He has not only pushed me to complete some of my many projects but has been an invaluable help in finalizing them. Thank you Mitch, I hope that this book will help many teachers to show their students the fun in learning about the Weather and the instruments used to record the various Physical characteristics used in the study of Meteorology.

TABLE OF CONTENTS

1. Introduction

It is really ironic that my first teaching position included four sections of Earth Science and one section of Physical science. I say ironic because first, my field was Biology and second, later in my career I learned how enthusiastic children really were about the weather. I never got around to teaching the unit on the weather that first year as I was promoted to teaching Chemistry and Physics half way through the year, two subjects that I loved and actually felt more comfortable teaching, but later in my career I came back to the Weather and then with a great passion.

Back to those children, my teaching career has been long and varied with stints teaching High School, Junior High, College and Elementary. I know I did not place them in any order for the simple reason that the Elementary students are the ones who got me thinking about this great topic, "The Weather".

It was while teaching Kindergarten through fifth grade Hands On Science that I realized how much Science, History, Art and just about any field of study could be incorporated into a unit on The Weather.

For the past twenty years I have been giving workshops on how to build working weather instruments and have during that time increased the levels from Kindergarten though High School. During those workshops I also included all of the information and activities you will find on the following pages. I have tried to keep the activities simple and the costs as low as possible with hints on how to increase the levels for older students.

Although I have used film canisters for several activities they are by no means the only materials that can be used, they were incorporated from a brief experience in the photo finishing business when film canisters were everywhere and I felt a need to put them to use.

Finally, I believe that Weather is a topic that everyone is interested in learning about and having a little understanding of the Science involved will make that learning a little easier.

I 'cannot tell you how many times I had an Elementary teacher ask me how they could get more Science teaching into their day. I would look at their Social Studies Curriculum and go from there. It was easy in New Jersey because there is a great section on the different regions of New Jersey that work swell with a study of rocks then there are the explorers and a perfect segue into ocean currents. In other words it does not take much to find ways to integrate Science and other disciplines, but weather is definitely the easiest.

You can work math into graphing the components of air, the altitudes of clouds, reading and calibrating weather instruments and time

lines for the creation of different weather instruments. It is also fun to look at the time line to correlate with the explorers and to see what instruments they had at their disposal. I remember once seeing a picture of Christopher Columbus using a telescope, which had not been invented yet!

I have worked with Art, Computer, and Music teachers along with classroom teachers to develop lessons that had a Weather theme.

2. THE WEATHER

I have often been asked by students, "Why do they call the weatherperson a Meteorologist when the person who studies meteors is called an Astronomer?"

My reply is to explain that the word meteor is from a Greek word that means "high up", and that lightning and shooting stars (meteors) were considered to be the same. It is interesting to note that in 350 BC, Aristotle wrote a book titled <u>Meteorology</u>, which was about the weather. Aristotle is also considered the founder of meteorology. A Meteorologist once told me, that the word meant "unpredictable" but I have never been able to find any etymological evidence for that connection, although I must admit that it would really make sense. We will look at the History of Meteorology more closely later on and we will also learn about the invention of each of the weather instruments studied in the second part of the book, Making Weather Instruments.

The study of weather should really start with a study of air and that is where I wish to begin. I have written this book in two parts, the first is to get the big picture of weather and the second is to show you how to make weather instruments that really work.

So where do we begin with our study of air? First we should look at the components of air; the gases, Oxygen, Nitrogen and although a very small component, Carbon Dioxide, then there is water and finally particles like dust. I like to have my students carry out a series of simple activities that each describe some component of air, then have them explain which component they have discovered.

THE COMPONENTS OF AIR

An activity to show that air contains Oxygen requires a more sophisticated apparatus and is best demonstrated by the instructor.

You will need a tall cylinder (a large graduated cylinder works well), an apparatus to clamp the cylinder upside down, a container to hold water, a large cork and a candle and match.

Clamp the cylinder upside down into the apparatus, place the candle on the large cork and float it in the container of water, light the candle. Slide the cylinder over the lit candle so the cylinder touches the water. After a few minutes the candle will use up all of the oxygen in the cylinder and you will see the water rise in the cylinder.

To demonstrate the presence of Carbon Dioxide in air you will need the indicator Bromthytmol Blue. If a student breathes through a straw into a water solution with the indicator it will turn yellow. The Carbon Dioxide changes the pH of the solution to an acid, Carbonic Acid.

There are really no easy ways to show students that air is mostly composed of the element Nitrogen so a little research would be a great way to let them find out. Have them prepare a circle graph to show the percentage of all of the gases in air.

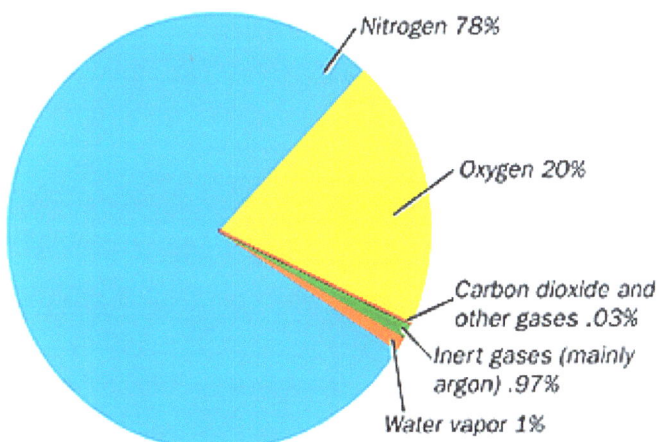

You can also have the students determine that air contains water by having them breathe on a mirror or just observing the condensation on the outside of a cold glass or soda bottle.

To show that air contains small particles like dust, you can have students look at sunlight streaming through a window and notice how the dust seems to stay suspended in the air. This is due to a phenomenon known as Brownian Movement and can also be used to show how air molecules are in constant motion; they are actually hitting the dust particles and keeping them up. There are several neat animations on the Internet to show Brownian Motion.

3. THE PROPERTIES OF AIR

Now that we have shown the components of air let the students experience some of the properties of air through the following activities. I have chosen; Air pressure (which is due to the mass of air), temperature, humidity (water content), volume and density to be explored by the students.

Activities to show that air exerts pressure and therefore has mass can include the following; the easiest way to do this is with a vacuum pump, an empty two liter soda bottle and a one holed stopper that fits the opening in the soda bottle. When the air is removed the soda bottle will be flattened.

You can also crush a can by placing a small amount of water in the can, then heating the can to boil the water. If you quickly cool the can in ice water the steam will condense and lower the volume inside the can thus allowing the air pressure to crush it. This activity requires tongs to move the can and should only be done as a demonstration by the instructor.

Another simple activity is to use a small suction cup. The reason that the suction cup sticks is because the air is pushing the suction cup down.

There are several other activities but I believe these two are the simplest and most effective methods to demonstrate that air exerts a pressure and therefore has a mass.

To determine that air takes up space (has volume), tape a small piece of paper towel on the inside bottom of a plastic cup (a transparent cup is best). The student is then asked to push the cup upside down into a container of water after predicting what would happen to the paper towel. This activity will show the student that air takes up space because the paper towel remains dry.

Although you may think that all students realize that air has temperature it is fun to see how they react to placing a cover (a small piece of cloth) over the bottom of a thermometer. Most will predict that the temperature will rise!

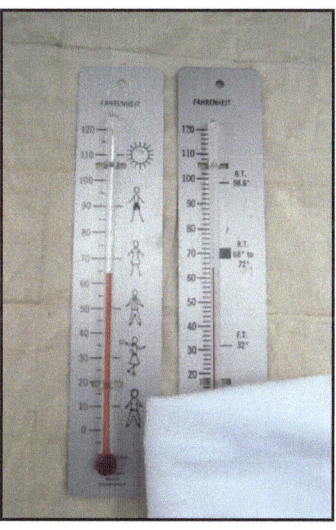

Another fun activity is to place a wet cover over the bottom of the thermometer and to have the students predict what will happen to the temperature. Again, they will be surprised to see that the temperature goes down due to evaporation.

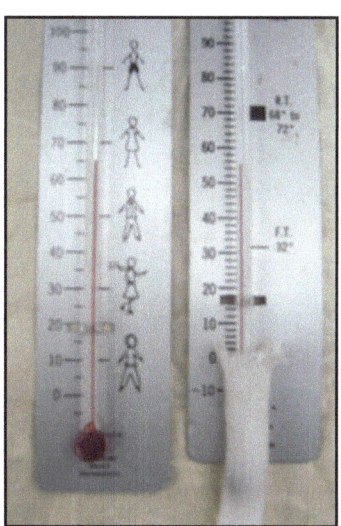

We have already seen that a simple demonstration to show the condensation of water using a glass of ice water will show that there is water in the air.

To determine the density of air is a little more difficult and will require; a 60cc hypodermic needle, a small amount of modeling clay, a nail, a drill, and an analytic balance.

First you have to pull the plunger out to read exactly 50 cc of air inside the hypodermic cylinder, then drill a hole through the plastic of the plunger at the top of the cylinder big enough to have the nail fit inside the hole. Place the nail in the hole and the small piece of modeling clay over the hole at the end of the hypodermic syringe. Find the mass of all parts and record as mass#1.

Remove the nail and modeling clay, push the plunger all of the way to the bottom. Place the modeling clay in the hole at the bottom of the syringe, pull the plunger back (it will not be very easy) and place the nail through the hole in the plunger. Find the new mass of all of the parts and record as mass#2.

Take the difference of the two masses and you will now have the mass of 50 cc of air, calculate the density.

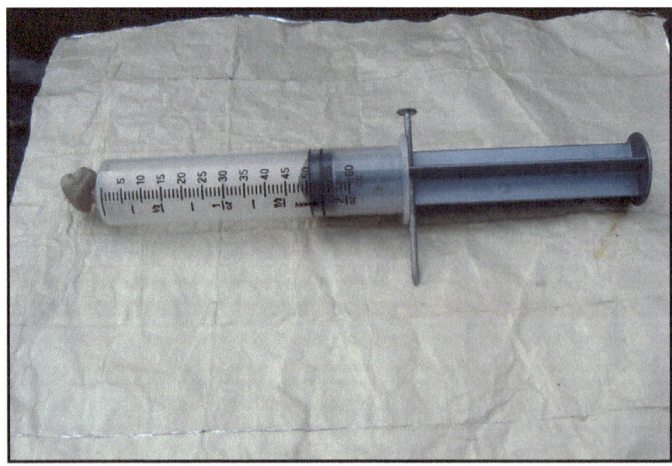

4. THE AIR AND THE ATMOSPHERE

Now that the students have knowledge of the components and properties of air they can now learn about how air acts in the atmosphere.

It would be fun to have the students get a weather balloon and a remote temperature sensor to see how the temperature of the air changes with altitude but if that is not possible then a little research will produce a picture of the different layers of air above the surface of the earth and the temperature and other properties associated with each. This is a good time to have them produce a scale drawing of the layers of the atmosphere.

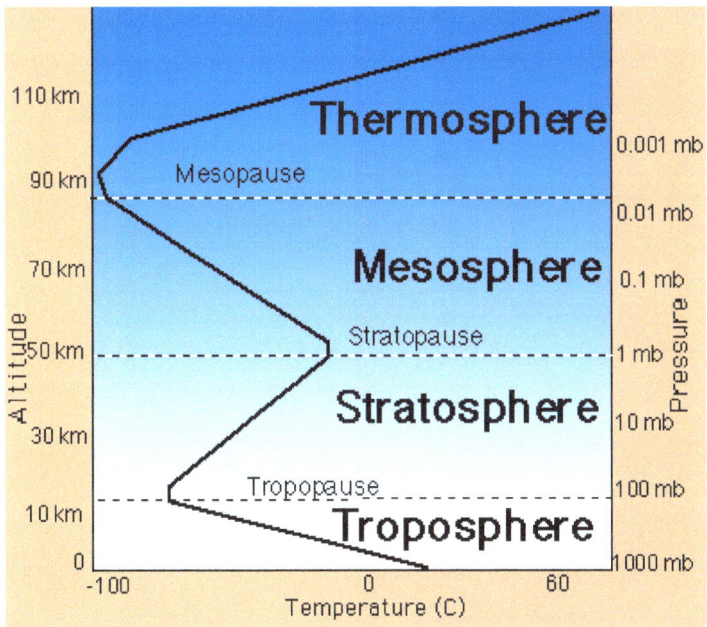

After they have learned the layers and that weather takes place in the layer known as the troposphere they can now move onto the study of the movement of air.

There are many different causes of wind. The first is the rotation of the earth, the Coriolis Effect, which is actually only a player over large distances and times. The Coriolis Effect can be demonstrated using a spinning disc and a pencil. Have one student spin a disc on a pencil while another student takes a second pencil and tries to draw a straight line from the edge to the center. A great animation can be found at: http://www.classzone.com/books/earth_science/terc/content/visu alizations/es1904/es1904page01.cfm?chapter_no=visualization

By the way the direction of water going down a drain is NOT due to the Coriolis Effect, it is much too small an area, it is actually due to the physical shape of the area around the drain.

A more influential cause of wind is the difference in temperature between the equator and the poles most notably the spawning of hurricanes off the coast of Africa at the equator and the movement of those hurricanes West and North following ocean currents.

Wind can be more easily understood to be the result of the difference in air pressure in the atmosphere. The difference in pressure is the result of differences in the temperature of the air with cold air being denser, and the warm air being less dense, thus resulting in a pressure difference.

Students may be most familiar with the phenomenon of sea breezes at the shore. During the day the land gets warmer than the water causing the air above the land to become less dense (lower pressure) and rising, this air is replaced by the cooler (more dense) air over the ocean, producing the sea breeze during the day. At night the water is warmer than the land and the breeze is toward the water, a good thing to know when flying kites at the shore.

Students may also be familiar with the winds that are associated with two weather fronts coming together A weather front is when a mass of either cold air or warm air comes in contact with a mass of air with an opposite temperature.

Thus a Cold front occurs when a mass of cold air confronts a mass of warmer air. The cold air being denser tries to slide under the warm air and at the place where the two masses meet we have a weather occurrence. The warm air rises above the boundary producing an area of low pressure and thunderstorms if the air has sufficient moisture. This weather usually does not last long but may be accompanied by strong winds. The symbol for a Cold Front on a weather map is a line with triangles on the bottom.

A Warm Front is when a mass of warm air confronts a mass of cold air the warm air being less dense rides up over the cold air and the boundary may produce precipitation if there is sufficient moisture in the air. This precipitation may last several days but is not usually accompanied by strong winds. The symbol for a Warm Front is a line with half circles on the bottom.

When two fronts are of equal magnitude a Stationary Front can form which can bring extended periods of precipitation. The symbol for a Stationary Front is alternating triangles and half circles on a line. For children this is the best possible scenario in the winter as it may produce large quantities of snow.

Students also love to learn about the more serious side of wind, as in tornadoes and hurricanes. As I have already mentioned hurricanes (on the East Coast) are caused by temperature differences of the ocean water between the equator and the North. A hurricane is a vast area of low air pressure that rotates in a counter clockwise direction. Other names for a hurricane include, typhoon and cyclone in other parts of the globe. Hurricanes get their energy from the warm ocean water and therefore lose energy once they hit a landmass.

Tornadoes are also the result of low pressure but this time they are affiliated with cumulonimbus clouds over land. They get their energy from the temperature differences inside the cumulonimbus cloud.

SNOW

Speaking of snow, children may be amazed to learn that there are many different types of snowflakes, in fact there are several different systems used to classify the shapes of snowflakes. The actual shape of a snowflake is determined by the humidity and temperature of the air, but temperature is the most important. One of the systems actually breaks the

shape of snowflakes into 35 different types. Johannes Kepler wrote the first scientific treatise on snow crystals: "Strena Seu de Nive Sexangula (A New Year's Gift of Hexagonal Snow)", although I have not read it a little research would be fun just to see how snowflakes can be different and not the old cliché, that all snowflakes have six sides.

Now, there is another old saying that the Innuit (the Eskimos) had many different words for snow. The odd thing is that the Inuit had only one word for snow but the Laplanders, another Northern people, did have many words for snow.

Snow is actually very interesting to study. Now that we know that there are many versions of snow we should understand that snow starts as super-cooled water in a cloud, this water may remain a liquid or form a crystal depending upon temperature and the availability of a nucleus for the water to condense on. The shape and size of the crystal will also vary with these factors.

While we are talking about snow this is a good time to talk

about other types of precipitation, especially hail. The formation of hailstones inside a cumulonimbus clouds is an interesting study of Physics and Meteorology. The vertical winds inside the cloud can lift a very large hailstone upwards until gravity takes over, with this process being repeated until gravity finally wins or the energy needed to maintain the vertical movement decreases. Students can see that a hailstone has layers and may assume that the number of layers corresponds to the number of times the hailstone has moved through the cloud. This is only partially correct, it turns out that the interior of the cloud is also composed of areas of differing moisture content that will also cause the appearance of layers.

5. WEATHER LORE AND WEATHER IDIOMS

WEATHER LORE

It is interesting to think how observant people are when it comes to the weather. There are many clues that can help the average person make short-term weather forecasts, such as the type of clouds in the sky, or even the wind direction and these have been used successfully for thousands of years. Children were taught about weather forecasting through poetry. You may be familiar with the following poem.

> *Red sun in the morning,*
>
> *Sailors take warning.*
>
> *Red sun at night,*
>
> *Sailors delight.*

The color of the sun is the result of the refraction of the suns rays by the atmosphere. Weather moves from West to East in the Northern Latitudes so a coming storm will have clouds, may be indicated by high Sirius clouds in the West, causing the red color as the sun shines on them. At night the sun would be shining on clouds to the East, indicating the storm has past.

`The problem with this is that any accumulation of dust or smoke in the atmosphere will produce red skies, so if a volcano had erupted thousands of miles to the West the sky would be red, the same would be true for a forest fire to the West.

When the wind is blowing in the North

No fisherman should set forth,

When the wind is blowing in the East,

'Tis not fit for man nor beast,

When the wind is blowing in the South

It brings the food over the fish's mouth,

When the wind is blowing in the West,

That is when the fishing's best!

Here we have Weather Forecasting by the direction of the wind. So, a Wind from the North would be cold, from the East usually means a storm, from the South would be warm, and from the West usually means good weather.

There are many other poems that were used in the past and it would be fun to have students research them for their accuracy as weather forecasting tools.

Great website:
http://weatherstories.ssec.wisc.edu/sayings/sayings.html

WEATHER IDIOMS

It is amazing how many times the weather comes up in our daily language. A great exercise would be to have students find a Weather Idiom and then explain how it relates to the weather. Some examples would be;

In the Doldrums

Under the Weather

It's a breeze

Don't steal his thunder

Every cloud has a silver lining

6. MISCONCEPTIONS ABOUT THE WEATHER

Children and adults have many misconceptions about the weather. One of the most common is that lightning never strikes twice in the same spot. The formation of lightning is very complicated but to simplify the process lightning it is a flow of electricity and this flow follows a path. Once that path has been created lightning will travel the path more than once as long as the storm that it producing the lightning is not moving too fast. The other factor to be considered is the distance from the source of the lightning to the ground, lightning is very fickle and although you may think being tall is bad it is not always the way it plays out, but you are still safer away from a tree during a lightning storm than under it.

THE HISTORY OF METEOROLOGY

I think that is safe to say that Meteorology may be the oldest of the sciences. I am sure that early man was aware of clouds and the weather that was associated with each type, even as far as saying that the appearance of Sirius clouds meant a change in the weather. They were also aware of wind direction and the weather associated with changes in the direction of the wind.

We do know that the Native Americans had many stories that they used to forecast the weather.

"If a muskrat builds his house toward the edge of the lake it means we will have a mild winter."

Logic: A muskrat needs open water to get out of his house. So if he builds near the edge of the lake, it means he knows that there won't be a long hard freeze.

We also know that as early as 340 BC Aristotle had written a book on Weather and that his pupil Theophrastus also wrote a book on forecasting the weather.

In 500 AD there was a book written on how clouds are formed in India.

In 1743, Ben Franklin observed that northeast storms begin in the southwest from correspondence with a friend in the Southwest.

7. HIGH AND LOW AIR PRESSURE MODELS

INTRODUCTION

Weather can be fair or stormy. Generally, fair weather is associated with high surface air pressure while stormy weather is associated with low surface air pressure. Broad-scale areas of high and low surface pressure dominate weather in middle latitudes and are simply called Highs and Lows.

Highs and Lows are regions where air pressures are higher or lower compared to the surrounding areas and are typically hundreds, or even thousands, of kilometers across. On a weather map, a large "High" or H symbolizes the location of highest pressure in a High whereas a large "Low" or L symbolizes the position of lowest pressure in a Low. Highs and Lows generally travel from west to east while exhibiting at least some motion toward the north or south. As they travel, they bring changes in the weather to the places along their paths.

This activity investigates (1) the horizontal and vertical air motions in Highs and Lows, and (2) the impacts of these motions on weather at locations under the influence of Highs and Lows.

PROCEDURE: CONSTRUCTION OF A MODEL HIGH PRESSURE SYSTEM

1. Using a copy of the map of North America found in Appendix find the H near St. Louis representing the center of a broad high pressure area. Lightly draw a circle on the map about 3 cm in diameter around the "H."

2. Place the map flat on your desk. If possible, stand up. (This exercise works better standing up.) Bring the thumb and fingertips of your left hand (if you are right-handed) or your right hand (if your are left-handed) close together and place them on the circle you drew.

3. Rotate your hand slowly clockwise, as seen from above, and gradually spread out your thumb and fingertips as your hand turns. Do not rotate the map. Practice this until you achieve as full a twist as you can comfortably.

4. Place your thumb and fingertips back in your starting position on the circle. Mark and label the positions of your thumb and fingertips 1, 2, 3, 4, and 5, respectively.

5. Slowly rotate your hand clockwise while gradually spreading your thumb and fingertips. Go through about a quarter of your twisting motion. Stop, mark, and label the positions of your thumb and fingertips on the map. Follow the same procedure in quarter steps until you complete your full twist.

6. Connect the successive dots for each finger and your thumb using a smooth curved line. Place arrowheads on the lines to show the directions your thumb and fingertips moved.

7. The spirals represent the general flow of surface air that occurs in a typical high pressure system (or High)

PROCEDURE: CONSTRUCTION OF A MODEL LOW PRESSURE SYSTEM

1. Using another copy of the map of North America found in Appendix, find the "L" near St. Louis representing the center of a broad low pressure area. Lightly draw on the map a circle about 3 cm in diameter around the "L".

2. Again, if possible, stand up. Place your non-writing hand

flat on the map with your palm covering the circle and your fingers and thumb spread out.

3. Practice rotating your hand counterclockwise as seen from above while gradually pulling in your thumb and fingertips as your hand turns until they touch the circle. Do not rotate the map. Practice until you achieve a maximum twist with ease.

4. Place your hand back in the spread position on the map. Mark and label the positions of your thumb and fingertips 1, 2, 3, 4, and 5, respectively.

5. Slowly rotate your hand counterclockwise while gradually drawing in your thumb and fingertips. Stopping after quarter turns, mark and label the positions of your thumb and fingertips. Continue the twist until your thumb and fingertips are on the circle.

6. Connect the successive dots for each finger and your thumb using a smooth curved line. Place arrowheads on the lines to show the directions your thumb and fingertips moved.

7. The spirals represent the general flow of surface air that occurs in a typical low pressure system (or Low).

Investigations: Characteristics of High & Low Pressure Systems

NOTES:

In the "High Pressure" model the students are bringing their hand down and clockwise. The down motion signifies that the air is cooler, drier and more dense, while the clockwise motion is the direction of the wind. It should be noted here that there is a common misconception about the density of wet and dry air. Most individuals have the idea that air that is wet is necessarily heavier than dry air. A little review of basic Chemistry is needed to explain the relative densities of wet and dry air. Humid air contains a large amount of water which has the molecular mass of 18 (2 Hydrogen at 1 each plus 1 Oxygen at 16), this water is displacing Nitrogen which has a molecular mass of 28 (2 Nitrogen's at 14). So the wetter air is actually lighter than the drier air. Ask any pilot and they will tell you that you get more lift in drier air than wet air.

In the "Low Pressure" model the students are bring their hand up and counter-clockwise. The up motion signifies that the air is warmer, moister and less dense, while the counter-clockwise motion is the direction of the wind.

CLOUDS

DESCRIPTION

OK, clouds are technically not weather instruments, but they sure are handy for both short and long term forecasting. I am sure it is safe to say that you can go back a long way and still find early man using clouds for their weather information. Even today most people are aware of the basic cloud types. They may not know their names but they sure know that those gray, low clouds usually mean rain. Or, that huge dark, ominous thing is most likely a thunderstorm in the making!

There are many, many different names for clouds, so we will keep it as simple as possible.

We will use Luke Howard's (1803) system. First, we will look at the shapes of clouds:

- **Cirrus** are curly or wispy

- **Stratus** are layered or flat

- **Cumulus** are round or puffy

Next we will look at the height of the clouds:

- **Cirro** is a prefix that means the clouds are high, above 10 km (20,000 ft)

- **Alto** is a prefix for middle height clouds with their bases around 6 km (6,000 ft)

- **Nimbo** at the beginning or nimbus at the end means the cloud is producing precipitation.

Now, you must also know that you can combine names to describe a cloud. So a Cirrocumulus cloud is a high cloud that is lumpy.

Some general rules of thumb are; Cumulus clouds usually mean nice weather but can build into Cumulonimbus clouds, which mean thunderstorms. Stratus clouds usually block most of the sun and can form into nimbostratus clouds, which means rain. Cirrus clouds, those highflying wispy clouds usually indicate a change in the weather for the worst.

Low clouds contain water droplets and high clouds contain ice crystals.

HISTORY OF CLOUD CLASSIFICATION

Past – People generally associated flat gray clouds with precipitation, and big tall clouds with thunderstorms.

1802 – Jean Baptiste Lamarc grouped clouds as; hazy, massed, dappled, broom like and grouped.

1803 – Luke Howard gave us; stratus, cumulus, cirrus, nimbus and alto, the names and prefixes we use today.

1887 – Ralph Abercromby and Hugo Hildebrandson revised the system by altitude and form.

1896 – Leon Teisserenc de Bort financed the publication of the first International Cloud Atlas.

See Appendices F & G for more information on different

types of clouds.

8. A CLOUD IN A BOTTLE

Clouds are really what weather is all about!

MATERIALS:

1. 2 or 3 liter clear soda bottle with cap.

2. A few drops of water.

3. A match or any aerosol.

4. A thermometer or temperature strip.

DIRECTIONS:

1. Place the thermometer, or better yet a temperature strip, in the empty soda bottle. Place the cap on tightly.

2. Record the temperature inside the bottle.

3. Squeeze the bottle with two hands and observe the temperature. It will go up.

4. Release the bottle and the temperature will return to normal.

5. Remove the cap and add a few drops of water, swirl around to wet the inside of the bottle.

6. Light a match and let it burn a few seconds and then drop it into the wet bottle, quickly place the cap on tightly.

7. Squeeze the bottle tightly, and then release quickly. You will see a cloud inside the bottle. Squeeze and it disappears, release and it appears!

WHY?

Squeezing the bottle increases pressure and temperature, also decreasing the volume, all of which you might have learned in High School Chemistry or Physics when you studied the Gas Laws. See there really was a reason to learn that little formula! So when you release the bottle the gas expands, the pressure drops and the temperature drops. The smoke supplies a surface for the water, which was vaporized when the temperature went up, to condense upon, thus forming a cloud.

SCIENTIFIC CONCEPTS

The Combined Gas Law states: $P_1V_1 / T_1 = P_2V_2 / T_2$ where P_1, V_1, and T_1 are the beginning pressure, volume and temperature. When you squeeze the bottle; P goes up, V goes down and T goes up. When you release the bottle; P goes down, V goes up and T goes down.

The net effect being a drop in temperature. Clouds need three things to form; water, a nucleus (dust, smoke, etc.) and a low temperature. Water comes from evaporation, the nucleus is dust that is normally in the air and the low temperature comes from many sources. It could be from the expansion of air as it rises, a cold front, altitude, etc.

9. A WEATHER/CLOUD CHART

Younger children love to keep track of the weather and this is a great way for them to keep track and learn some new Meteorological terms.

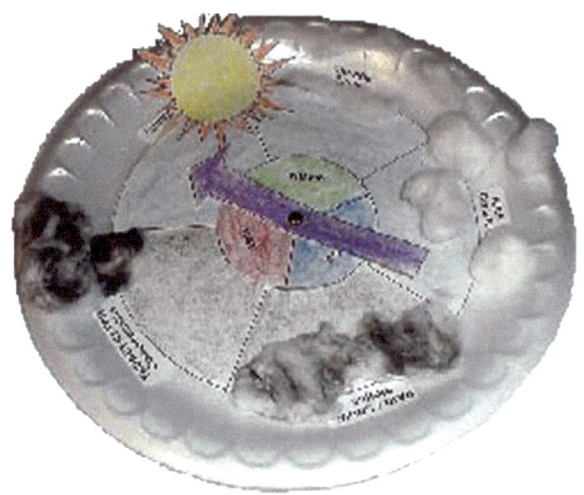

MATERIALS:

1. Pages of templates (Appendix)

2. Cotton balls or cotton batting.

3. Glue stick or other adhesive.

4. Brass fastener (1 cm)

5. Hole punch (preferably a rotary punch)

DIRECTIONS:

1. Color the sections that say sunny and nice, light blue, all other sections are colored gray.

2. Color the sun yellow and orange.

3. Color; Hot, red; cold, blue; warm, green.

4. Color the pointer any color except one that has been used.

5. Cut out the two circles and the pointer. You may want to paste the large circle on a large paper plate.

6. Assemble the chart; pointer on top of small circle on top of the large circle. Fasten with the brass fastener.

7. Use several cotton balls to represent Cumulus clouds and paste them on the section that says nice (Cumulus).

8. Take several cotton balls and stretch them into thin clouds. Paste them on the section that says change (Cirrus).

9. Take a cotton ball and rub it on a piece of paper that has black crayon to make the cotton gray. Stretch it flat and paste it on the section that says rain I snow (Stratus).

10. Take two cotton balls and rub on the black crayon to get them almost black. Paste the balls one above the other on the section that says thunderstorm (Cumulonimbus).

11. Paste the sun on the section that says sunny.

WHY?

Children can learn the names of clouds by using them. They can learn how clouds can be used to forecast the weather.

SCIENTIFIC CONCEPTS:

Cumulus clouds are formed when warm air rises and then condenses on particles in the sky, thus they indicate nice weather. Cirrus clouds are parts of higher clouds that have been blown off in front of a front, they indicate a change in the weather, usually rain. Stratus clouds are gray because the hold a large amount of water, they are low to the earth and usually mean rain or snow. Cumulonimbus clouds are tall clouds produced by vertical movement of air. They can form quickly and may have heavy precipitation and lightning.

10. THERMOMETERS

DESCRIPTION

One of the most common aspects of weather is the idea of temperature. Temperature by definition is a measurement of the average Kinetic Energy of molecules. Which means that as molecules go faster they have more energy and thus a higher temperature. Temperature and heat are not measurements of the same thing. Heat is a measurement of the Total amount of Kinetic Energy in a body and is therefore a function of the average Kinetic Energy, the type of material and the amount of the material. Thermometers are instruments that measure Temperature and use units like Fahrenheit, Celsius or Kelvin degrees.. To measure heat you would use an instrument called a calorimeter and you would use calories, kilocalories or BTU's.

There are many types of thermometers that have been developed over the years (see the following section, The History of Thermometers). The glass thermometer is the most common. It works when a liquid such as alcohol (Mercury is no longer used because of health dangers) expands as it is heated. Many books talk about building a water thermometer which is simply a straw in a stoppered container filled with colored water. The problem is that there is only a slight expansion of the water and of course the problem with spilling because the tube is not sealed as in a commercial thermometer. There are also air thermometers but these are also subject to changes in atmospheric pressure.

It is for these reasons that I favor the thermometer on the

following pages. It is safe, not messy, and fairly sensitive to changes in temperature. Not to mention, easy and inexpensive to construct. In fact, with a little help can be constructed by children in first or second grade.

HISTORY OF THE THERMOMETER

Nature – Snowflakes have been telling the temperature by their shape since, well since there has been snow. Crickets have been telling the temperature for, well almost as long as there have been crickets. See thermometers for the how?

62 AD – Heron of Alexandria described several complex devices to measure temperature in his *Pneumatica.*

1592 – Thermoscope, Galileo, used the expansion of air to indicate temperature change, the air pushed down on alcohol when it expanded, others were more complex mechanical devices, scales were general not numerical.

1611 – Sanctorius Sanctorius (a colleague of Galileo), devised a scale using melting snow and the heat from a candle, into 110 equal parts.

1624 – Leurechon first used the name thermometer.

1632 – John Rey, first liquid thermometer using water.

1644 – Grand Duke Ferdinand increased the sensitivity by hermetically sealing the end of the tube. Ferdinand also created a group of artisans to produce thermometers that were consistent., they also were the first to try Mercury in their thermometers, although they preferred alcohol because it had a greater expansion coefficient. Amazingly, some of these thermometers are still in existence.

1708 – Ole Romer, developed a scale using $60°$ for the BP of water and $7.5°$ for the FP of water, these were eventually changed, using the temperature of blood as the upper limit, it being 22.5°.

1700-30 – Daniel Gabriel Fahrenheit used 32° for the MP of ice and 96° the temperature of blood, this then gave 212° for the BP of water.

1730 – Rene-Antoine Ferchault Reaumur made a thermometer using a diluted alcohol mixture, he set 0° for the FP of water and calibrated the thermometer so the BP of the alcohol was 80°, which would have given a value of 100° for the BP of water. But, because of confusion in the understanding of how the thermometer was actually calibrated it was assumed that the 80° mark was the BP of water, so Reaumur although actually having developed a metric thermometer was not credited with doing so.

1742 – Anders Celsius developed a thermometer that took into account the fact that the BP of water was not only dependent upon the temperature but the atmospheric pressure.

1848 – William Thomson (later known as Lord Kelvin) combined the work of Robert Boyle, Jacques-Alexandre-Cesar Charles, Joseph-Louis Gay-Lussac and Sadi Carnot to develop the Kelvin scale, which basically states that at a temperature of -273.15C° all matter is in either a liquid or solid state because the molecules are packed as tightly as possible and for all practical purposes occupy the smallest volume. In other words they are almost stopped.

1821 – Sir Humphrey Davy determined that the electrical resistance of a material is dependent upon temperature. He built an electrical resistance thermometer using platinum and it is one of the most stable and accurate thermometers available.

1822 – Thomas Johann Seebeck built a thermocouple, which uses two dissimilar metals (Cu and Fe) to create a current when heated.

OTHER METHODS OF DETERMINING TEMPERATURE

a.   Acoustic – the velocity of sound through a medium is dependent upon the temperature of the medium, so by determining the velocity of sound you can determine the temperature.

b.   Magnetic – accurate at very low temperatures (1K) these thermometers employ the temperature dependency of the magnetic susceptibility of materials.

c.   Thermal noise – the random motion of electrons in conductors produce static, this static is also a function of temperature.

d.   Pyroelectric – some solids exhibit spontaneous polarization (they switch charges) when heated.

e.   Temperature indicators – certain materials change state at a single temperature, e.g. crayons, and this can be used to determine a certain temperature.

f.   Thermography – temperature mapping using infrared light.

g.   Thermistors – diodes that change resistance with temperature.

h.   Bimetallic strips – different metals, expand (and contract) at different rates causing a strip to bend, the degree of bending can be used to determine temperature.

11. A THERMOMETER

Most thermometers employ the expansion of a liquid; Mercury or Alcohol. Not this one!

MATERIALS:

1. One film canister any cylindrical type
2. Aluminum foil – heavy duty
3. Masking tape
4. 3"x5" index card
5. Scissors
6. Protractor

DIRECTIONS:

1. Cut a piece of the Aluminum foil the width of the masking tape and 8cm long.

2. Tape a piece of masking tape the entire length of the foil.

3. Trim any excess tape or foil and cut the foil down the center, this will be the pointer.

4. Place the protractor on the White side of the index card perpendicular to the card in a position so the arched part of the protractor is about 0.5 cm from the edge. Trace the curved part of the protractor on the card.

5. Remove the cap from the canister, and cut a slit about 1cm long down the side of the canister. Note, any similar container will work.

6. Place one end of the pointer in the slit so the pointer is perpendicular to the canister. Replace the cap.

7. Make a loop of masking tape and place it on the cap.

8. Place the cap on the index card and center so the pointer just passes the curved line on the card.

9. Straighten the pointer and mark the card RT (room temperature) where the pointer crosses the line.

10. Place the Thermometer in the freezer and observe in one hour, mark the card FT (freezing temperature) where the pointer crosses the line.

11. Place the thermometer under a lamp for one hour and mark the card HT (high temperature) where the pointer crosses the line.

12. If you want you can find the precise temperatures for each position and develop your own temperature scale or use one of the present s scales.

WHY?

As the Pointer gets colder the Aluminum contracts more than the masking tape and pull the pointer to the side with the Aluminum. When the pointer gets warm, the Aluminum expands more than t he masking tape and pushes the pointer away from the side with the Aluminum.

SCIENTIFIC CONCEPTS:

Solids like Aluminum expand when heated and contract when cooled because the molecules speed up and require more space or slow down and require less space. They will continue to speed up and expand until the molecules are going so fast that the material becomes a liquid or the molecules escape into the air as a gas.

12. BAROMETER

DESCRIPTION

Of all of the properties of weather the most important but least understood is air pressure. It is the pressure of the air that makes a beautiful cloudless day so perfect or a rainy day so, well so imperfect. It is also air pressure that makes our ears pop on an elevator or a mountain road, the vacuum cleaner pick up dirt (no the dirt is not sucked in), and the soda go up the straw.

The air above us, consisting of matter in the form of a gas has mass. The mass of the air is pushing on the surface of the earth and causes air pressure.

Although it is quite high about 1 Newtonl cm^2, or 14.7 pounds/in^2, we do not notice it because it is pushing down, up, and sideways at the same time. It is only when there is a difference in the air pressure that we notice. Like when you lower the pressure in your mouth and the soda is pushed up the straw by the normal atmospheric air pressure. Air pressure is responsible for much of the weather that we have come to know and hate.

The meteorologist talks about Low's and High's. A low pressure system has a lower atmospheric pressure and winds that usually rotate counterclockwise in the Northern Hemisphere. It is also associated with bad weather, and in extreme cases can cause hurricanes. A high pressure system has high atmospheric pressures and winds that usually rotate clockwise in the Northern Hemisphere. They are associated with clear skies and generally nice weather.

The instrument used to measure air pressure is the barometer. It measures air pressure in many different units. In the US, air pressure is normally measured in "inches of Mercury". This means that normal air pressure would push mercury up a closed tube to a height of 30 inches. It would also push water up a closed tube to a height of about 30 feet! In the metric system we use millimeters of mercury or more correctly Pascals, but those are for another time and place. All barometers of this type; a column of mercury in an open dish are not the safest things to have around. So we use barometers that use a difference in air pressure, called aneroid barometers for safety and to save space. In fact the barometer that you will be building is a very simple and safe aneroid barometer.

HISTORY OF THE BAROMETER

4th century BC – Aristotle "Nature abhors a vacuum".

1638 – Galileo showed a vacuum can exist but credited the idea to the cohesive nature of water.

1641 – Gasparo Berti built an apparatus to prove that a vacuum does not transmit sound.

1644 – Evangelista Torricelli (a friend of Galileo) , Florence, Italy, had his assistant Vincenzo Viviani builds a Mercury barometer.

1648 – Florin Perier (brother in-law of Blise Pascal) performed an experiment for Pascal that showed the atmospheric pressure was lower on the top of a mountain than the bottom

1657 – Otto von Guericke performed the famous experiment with the sphere that had been evacuated being pulled apart by two teams of four horses each to demonstrate the tremendous force of the atmosphere.

17th century – Thunder Glass, Netherlands

1698 – Gottfreid Leibniz described an aneroid barometer.

1843 – Lucien Vidie built a satisfactory aneroid barometer.

13 A THUNDER GLASS

The Thunder Glass, or water barometer dates back to early 17th century Netherlands.

MATERIALS:

1. A film canister, clear cylindrical with cap, preferably a cap that fastens inside.

2. Rotary punch, or hole punch.

3. A piece of clear plastic tubing about 12 cm long (tubing used for filters in aquariums works well because it is soft).

4. Water

5. Food coloring

6. Pattern to hold tube and canister.

DIRECTIONS:

1. Punch a hole just a little smaller than the diameter of the tubing as far down inside the canister as you can.

2. Place the tubing inside the hole so that it extends into the canister about 0.5 cm.

3. Fill the canister about 2/3 with water, add food coloring to the water.

4. Place the cap on the canister.

5. Make sure some water has risen up the tube if none has, add more.

6. Trace the pattern onto a piece of cardboard, cut out and slide over the canister and tube.

7. Watch the level of the water in the spout, to predict the weather.

WHY?

Changes in the air pressure cause the water to rise and fall in the spout. Low water level indicates high pressure and nice weather. High water levels (it may overflow) indicate low air pressure and Poor weather.

SCIENTIFIC CONCEPTS:

The air pressure inside the canister is constant (whatever it was when the Thunder Glass was constructed), as the air pressure outside changes it will either push the water back into the canister or allow the air pressure inside the canister to push the water up the spout

14. WEATHER VANES

The word "vane" actually comes from the Anglo-Saxon word "fane", meaning flag. Originally, fabric pennants would show the archers the direction of the wind. Later, the cloth flags were replaced by metal ones, decorated with the insignia or coat of arms of the lord or nobleman, and balanced to turn in the wind. From these antecedents come the banners, which the early American colonists favored for their meeting halls and public buildings.

A BRIEF HISTORY OF WEATHER VANES

The history of weather vanes is an interesting one, which spans many centuries and travels over many countries.

48 BC – Adronicus built a 4 to 8 ft tall weather vane to the god Triton

9th cent – quadrant shaped bronze weather vanes from Viking ships.

The pope decreed that every church show a cock on its dome

or steeple.

1040 – "Bronze Banner", Kirche zu Soderala, Halsingland, Sweden

11th cent – Bayeux Tapestry includes a scene of a craftsman attaching a rooster vane to the spire of the Westminster Abbey.

Middle Ages – It is probably the banners, which flew from

medieval towers in Britain, Normandy and Germany, which are the precursors to our modern weather vanes.

15. A WEATHER VANE

Weather vanes are easily the oldest weather instrument.

MATERIALS:

1.   A paper or Styrofoam plate (9")

2.   Plastic drinking straw, (approx 5mm diam.) (stripes help).

3.   Small plastic straw (approx2mm diam)

4.   Brass fastener (4cm, (1.5 inches) or longer)

5.   Crayons.

6.   Scissors

7.   Glue.

8.   Patterns for rooster, arrow and Compass Rose in Appendix C.

9.   One large bead, with a hole large enough for the brass fastener to fit through.

DIRECTIONS:

1. Color and cutout the rooster, arrow point and Compass Rose.

2. Fold and glue the Rooster (do not glue bottom) and Arrow point.

3. Make a slit in the front of t he straw for the arrow point, this is where the stripes help in aligning the top and bottom. Slide the arrow point in and staple or glue

4. Add glue to the bottom of the rooster, then glue to the outside of the opposite side of straw.

5. Punch a hole about 9 cm (31/2 inches) from one end of the straw perpendicular to the rooster.

6. Make a slit in the center of the paper plate for the brass fastener.

7. Place the fastener in the slit from the bottom.

8. Place the bead on the fastener.

9. Cut the thin (2mm) straw about 1 cm long. Place this straw in the hole you punched in the larger straw.

10. Place the larger straw on the fastener. Bend the fastener to hold.

11. Tape the fastener in place on the bottom.

See detail below

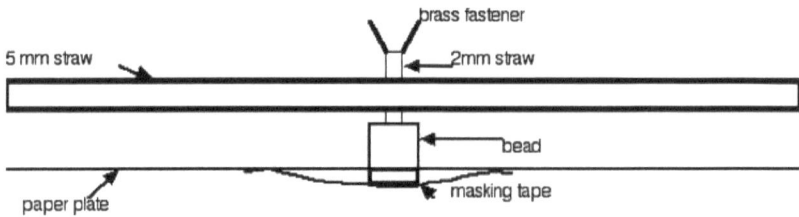

WHY?

The position of the rooster causes the arrow to always point in the direction from which the wind is blowing.

HISTORY OF THE ANEMOMETER

Nature – The blowing of grass, leaves, and trees has always been an indicator of how strong the wind blows and in fact was eventually translated into a quantitative scale by Beaufort.

1450 – Leon Batista Alberti made a swinging disk anemometer.

1664 – Robert Hooke and half a world away the Mayan Indians also created swinging disk anemometers. The Mayans called theirs a "wind tower".

1833 – E.P.Archibald flew anemometers on kites to measure the speed of the wind a various altitudes.

1850 – Thomas Romney Robinson built a 4-cup anemometer.

? William Henry Dines built a "pressure tube anemometer" sometime in the nineteenth century.

INSTRUMENTS THAT MEASURE WIND SPEED:

1. Anemometers
2. Aerovanes, use blades like an airplane propeller.
3. Radar
4. Satellites
5. Rawinsonde measurements, track drift of weather balloons.
6. Pitot tubes measure speed of airplanes.
7. Sonic anemometers use the speed of sound to measure instantaneous

16. AN ANEMOMETER

Anemometers come in many forms, this style although fun to build is more a model, to get a better appreciation for wind speed use the Beaufort scale.

MATERIALS:

1. A film canister, any round style.

2. Rotary punch, or hole punch.

3. 2 plastic drinking straws, about 20cm (8inches) long and 5 mm (1/4inch) in diameter. Make sure they are not the type that can bend easily!

4. Razor knife or other sharp knife.

5. Fine sand. (enough to fill the film canister)

6. One 8d nail (about 8 cm or 3 inches long)

7. Four small cups (the kind used in the bathroom work well, 6oz)

8. One large bead, with a hole large enough for the nail to fit

through. (You can make your own by taking a small section of straw about 1.5cm (1/2 inch) long and punching a hole through both sides in the center.

DIRECTIONS:

1. Punch a hole in the top of the film canister large enough for the 8d nail.

2. Make a slit in the center of one of the straws about 1 cm (1/2 inch) long

3. through both sides.

4. Slide the other straw through the slit until you have a perfect +.

5. Punch a small hole through the center of the two straws.

6. Punch a hole in the side (near the bottom) of a cup large enough for a straw , place another smaller hole opposite. Repeat for all 4 cups.

7. Slide one end of one of the straws that are crossed into the large hole in a cup and then work the end into the smaller hole. Repeat for all for cups, make sure all cups are facing the same direction.

8. Fill the canister with sand to the top, allow room for the top to snap on.

9. Place the nail through the crossed straws, into the bead and into the hole in the top of the canister.

10. Gently blow on the cups to see the anemometer rotate.

WHY?

Obviously, the force of the wind is converted to the rotary motion of the cups.

Practically, the shaft (the nail) would be attached to a generator and then to a meter, the meter would be calibrated in km/hr or miles/hr instead of volts. Other types of anemometers have a mechanical attachment to the shaft that causes an arm to go up along a calibrated device, and others use the Bernoulli effect to move a small ball up a calibrated cylinder.

17. AN ANEMOMETER #2

Anemometers come in many forms; this style is simple to build and works well.

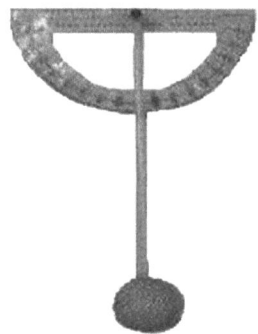

MATERIALS:

1. A protractor, or use the template in the Appendix (glue to cardboard to increase strength).

2. Rotary punch, or hole punch.

3. One ping-pong ball.

4. About 30 cm of very fine thread or fishing line.

5. Elmer's glue or equivalent.

6. Wind scale in the Appendix.

DIRECTIONS:

1. Glue one end of the thread to the ping-pong ball.

2. Attach the other end to the hole in the middle of the base of the protractor.

3. Invert the protractor so the ping-pong ball hangs straight down (90°).

4. When the wind blows the string will move away from 90°, count how many degrees and compare to the chart to determine the wind speed.

WHY?

Obviously, the force of the wind is converted to the horizontal movement of the ping-pong ball. The greater the force the greater the tendency for the pong-pong ball to be at 180o. The numbers on the chart were determined by plotting the angle against the wind speed from a calibrated anemometer and also the Beaufort scale (in Appendix).

18. HYGROMETER

DESCRIPTION

The water cycle is more correctly known as the hydrologic cycle, but in either case it is saying that the amount of water (all 1.42 x 1021 liters of it) is constant. It just does not stay in one place for any great length of time. Somewhere in our education the idea that the earth's surface is approximately 70% water was posed, and is for all practical purposes correct. The problem is more, where is the water? Of course, the oceans hold a good amount, and then there are rivers, lakes, streams, ponds, puddles, etc. And, oh yes, this is about the weather, how about the atmosphere? Incredibly the total percentage of water on the earth breaks out to roughly; 95.1% in the oceans, 4.9% as fresh water that is virtually inaccessible, being ice, ground water, in living things etc. and a measly 0.01% freshwater with only 15% of that in the atmosphere. Although this does not sound like much, we have to remember that in the course of a year the atmosphere actually releases 30 times that amount of water, which is considerable.

Now that I have impressed you with all of these numbers, let us take a look at how the water actually gets into the atmosphere. With the large surface area of the oceans, it is easy to see the evaporation is the primary source of water entering the air, this is augmented by geothermal release from volcanoes and hot springs and finally transpiration from plants, although I would bet that sublimation plays a rather significant roll in this cycle even if I have not been able to find confirmation! And then there is the remote possibility that we are getting water from outer space, in the form of small "snowballs". The other side of the cycle, the down side you might say, is precipitation in the form of; rain, snow, sleet, hail, etc.

Why all of the fuss about the water cycle? Well, to put it bluntly the water cycle is not only responsible for our weather it is our weather. A perfectly clear day with no clouds in view or an overcast day with rain falling in buckets both have water involved, it is only a matter of degree. The clear sky, although, cloudless is still holding a fair amount of water in the form of vapor, while the rainy sky has the atmosphere saturated with water, with a relative humidity of 100%.

MEASURING WATER IN THE AIR

The measurement of the water content of the air is called humidity and is on a scale of 0 to 100%.

The procedure for measuring humidity falls into several categories; the water vapor can be measured directly by some rather sophisticated instruments that employ infra red, reflection of light etc., it can be measured indirectly by comparing the temperature of the air to the temperature drop caused by evaporation as in most hygrometers or psychrometers, the humidity can be measured by the effect of water on something as in the protein in a human hair (the frizz effect), or it can be measured by the absorption of water as with the color change registered by Cobalt Chloride in those little weather houses.

Before we turn our attention to actually building a working Hygrometer we should keep in mind that the amount of water vapor in the air (the humidity) is a function of three things; the amount of water in the air, the temperature and the pressure. It is for these reasons that any hygrometer that is not designed with these in mind will not be very accurate, hence the inaccuracy of most hygrometers in home weather stations. The evaporation hygrometer that we will build takes into account the temperature but not air pressure (fortunately, pressure has the least effect on humidity). The hair hygrometer although very sensitive and capable of giving continuous immediate readings does not take into account either temperature or pressure.

Now to learn how to build two types of hygrometer; an evaporation hygrometer that can be built and understood by any child who can read a thermometer) and a hair (frizz effect) hygrometer which because of the difficulty in assembly, is best constructed by older children, at least 5th grade

HISTORY OF THE HYGROMETER

1450 – Nicolas Cryfts suggested the use of dried wool on scales to measure the water in the air.

– Leonardo Da Vinci probably built the first Hygrometer!

15th & 16th centuries – sponges were used to get a relative measure of the water content of the air !

1655 – Ferdinand II of Tuscany built a condensation hygrometer.

17th century – Wood, gut, string cord, hair and Gold beaters skin (ox-gut, which is still used today) were tried.

1751 – J.B. LeRoy built a dew-point hygrometer.

1755 – William Cullen (1710-1790) noticed that a thermometer with alcohol on it recorded a lower temperature than one that was wet with water. He concluded that the lower temperature was due to evaporation.

1783 – Horace B. deSaussure built a hair hygrometer (hair shrinks 2.5% between 100% and 0% humidity).

1785 – Seaweed was tried.

18th century – Evaporation of water was used (the advent of the psychrometer).

1820 – Samuel Daniel built a dew-point hygrometer.

Modern – Some of the more sophisticated dew-point hygrometers employ a beam of light on a precision mirror that can have its temperature changed by a thermoelectric device. When the temperature reaches the dew-point the beam of light is changed, The temperature of the mirror and the current to the thermoelectric device are recorded and used to determine the dew-point. The mirror can be rye heated and the cycle can be continued indefinitely, thus giving continuous readings.

19. MAKING A HYGROMETER

Hygrometers and psychrometers measure the amount of moisture in the air and can be used to determine the relative humidity of the air.

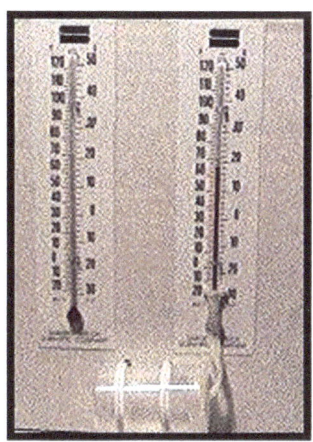

MATERIALS:

1. One film canister preferably clear.

2. Two small thermometers calibrated in C^o (F^o can be used but you will need the F^o conversion chart).

3. Length of heavy shoestring about 10 cm (4 inches)

4. Rotary punch

5. Conversion chart

6. Heavy cardboard

7. Several ties

DIRECTIONS:

1. Cut a piece of heavy cardboard large enough to hold both thermometers and the film canister below one of the thermometers.

2. Punch a small hole in the film canister lid.

3. Cut the shoelace and slide the shoelace over the bottom of one of the thermometers.

4. Fasten both thermometers to the cardboard using the ties.

5. Fasten the film canister below the thermometer with the shoelace.

6. Insert the remaining length of shoelace in the hole in the film canister.

7. Remove the lid of the canister and carefully, fill the canister with water, replace the lid.

8. Observe the temperatures on both thermometers until the thermometer with the sock STOPS getting lower. (If it is raining or a very damp day, the thermometer with the sock may not change).

9. Find the dry bulb thermometer reading on the right hand side of the chart.

10. Find the difference between the wet and dry-bulb thermometers and locate this number on the top of the chart, the intersection of the two numbers is the relative humidity. (EXAMPLE: the dry-bulb thermometer reads 24oC and the wet-bulb thermometer reads 20oC, the relative humidity is 69%)

WHY?

The sock will carry water to the thermometer by adhesion. The water will evaporate and cause the temperature to go down.

SCIENTIFIC CONCEPTS:

Heat is required to cause water to evaporate. The rate of evaporation is dependent upon the temperature and the amount of water already in the air. When air is saturated (has the most water it can hold at that temperature) the relative humidity is 100%

20. HYGROMETER #2

Hygrometers and psychrometers measure the amount of moisture in the air and can be used to determine the relative humidity of the air.

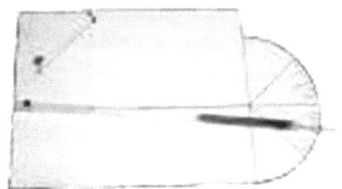

MATERIALS:

1. Template for Hygrometer #2 from Appendix.

2. One clean, dry human hair about 20 cm long.

3. One plastic drinking straw cut to 20 cm.

4. 3 straight pins or equivalent.

5. One toothpick.

6. Heavy cardboard, or equivalent.

7. Modeling clay.

8. Elmer's glue or equivalent.

DIRECTIONS:

1. Cut a piece of heavy cardboard large enough to glue hold the template, glue the template on the cardboard.

2. Place holes in the straw to correspond to the locations on the template for the fulcrum and the hair. (note, they will be 90° apart)

3. Place one end of the hair in the hole and glue in place.

4. When the glue has dried, place some modeling clay in the end of the straw (nearest hair) and place a toothpick in the clay to form a pointer.

5. Pin the straw to the template, make sure the straw moves freely.

6. Place the other pins in their correct positions.

7. Carefully, wrap the hair over the forward pin and tie to the pin in the back (note, once the hair has been tied you may want to move this pin backward or forward to have the pointer horizontal for the beginning of your data collection).

8. Place a 2 cm x 18 cm slit in the top of a 2 liter soda bottle to use as a protective case for your hygrometer.

9. Place a few small pieces of modeling clay on the bottom of your hygrometer and carefully insert into the soda bottle> (Note. you may have to cut the back of the pins off to get the hygrometer into the bottle, or make the slit wider)

10. Record the initial location of the pointer and watch the hygrometer as the weather changes. To calibrate your hygrometer bring it into the bathroom when you take a hot shower. Mark the location of the pointer to indicate 100 % humidity.

WHY?

Hair shrinks as the humidity decreases. Blond hair is the most sensitive to humidity. (Now you know why some people have "Bad hair days"). Hair is made from the protein keratin that is wound in a coil. Very weak bonds called "Hydrogen bonds" hold the turns of the coil together. These bonds are broken by water, thus causing the hair to get longer when wet. The bonds reform when the hair dries. [Note, frizzy hair is caused by water getting under the scales of the cuticle (the outer part of hair) and making the hair fatter]

SCIENTIFIC CONCEPTS:

As the hair changes length it moves the straw up and down. Because the straw is acting as a third class lever the pointer will move a greater distance, making it easier to see very slight changes in the length of the hair, and therefore the humidity.

21. RAIN GAUGES

Rain gauges are instruments that measure the amount of precipitation very accurately.

MATERIALS:

1. Any container that will hold water.
2. Ruler.

DIRECTIONS:

1. Place the container in a location so the rain gauge will be collecting only rain falling from the sky, make sure there are no obstructions.
2. After a rainfall use the ruler to measure the height of the water.

WHY?

Professional rain gauges are designed so the collecting tube is exactly 1/ 10 the area of the collecting funnel. This allows for very precise readings. For example, one tenth of an inch of rainfall will fill one inch of the collecting tube. Although it is not impossible to build an accurate rain gauge, it would be very difficult, so we will use a ruler to get a crude measurement.

HISTORY OF THE RAIN GAUGE

Nature – Probably the amount of water in puddles, streams, etc was used as an Indicator of rainfall.

400BC – A rain gauge is described in Arthastra by the Indian author Kautilya.

1441 – King Munjong, a Korean king invented a rain gauge to measure rainfall at the palace.

1639 – Benedetto Castelli described a rain gauge in a letter to Galileo Galilei.

1662 – Christopher Wren created a tipping bucket rain gauge.

22. REFERENCES

BOOKS

Be an Expert Weather Forecaster, Barbara Taylor-Clark

The Weather Book, Jack Williams

Hands-on Meteorology, Zbigniew Sorbjan

WEBSITES

American Meteorological Society

http://www.ametsoc.org/dstreme/junction/

Weather Net: http://cirrus.sprl.umich.edu/wxnet/

Today's Space Weather

: http://science.msfc.nasa.gov/newhome/virtualtour.htm

Atmospheric Pressure:
http://www.pbs.org/ktca/newtons/tryits/14/sciencetryits.html

http://ww2010.atmos.uiuc.edu/(Gh)/guides/mtr/prs/def.rxml

lightning: http://www.mos.org/learn_more/cheapbook

Weather instruments, History, etc: http://cicgi.msu.edu/~rtsmith

http://cicgi.msu.edu/meteorology

The Educator's Cheapbook:
http://www.mos.org/learn_more/cheapbook

Anemometers: http://www.sbsd.k12.ca.us/sbsd/specialprog/msb

Meteorology: The Sky's the Limit:
http://www.mcet.edu/meteorology

The Weather Project:
http://www.lelycee.org/projets_pedagogiques/mete

spacelink.nasa.go miscellaneous:

http://www.nxdc.com/weather/

http://www.srh.noaa.gov/tlh/wxhwy.html

http://weather.hypermart.net/

http://www.nws.noaa.gov/

http://www.earthwatch.com/

http://www.spaceweather.com/

http://www.weatherimages.org/

http://www.weatherman.com/

APPENDIX A

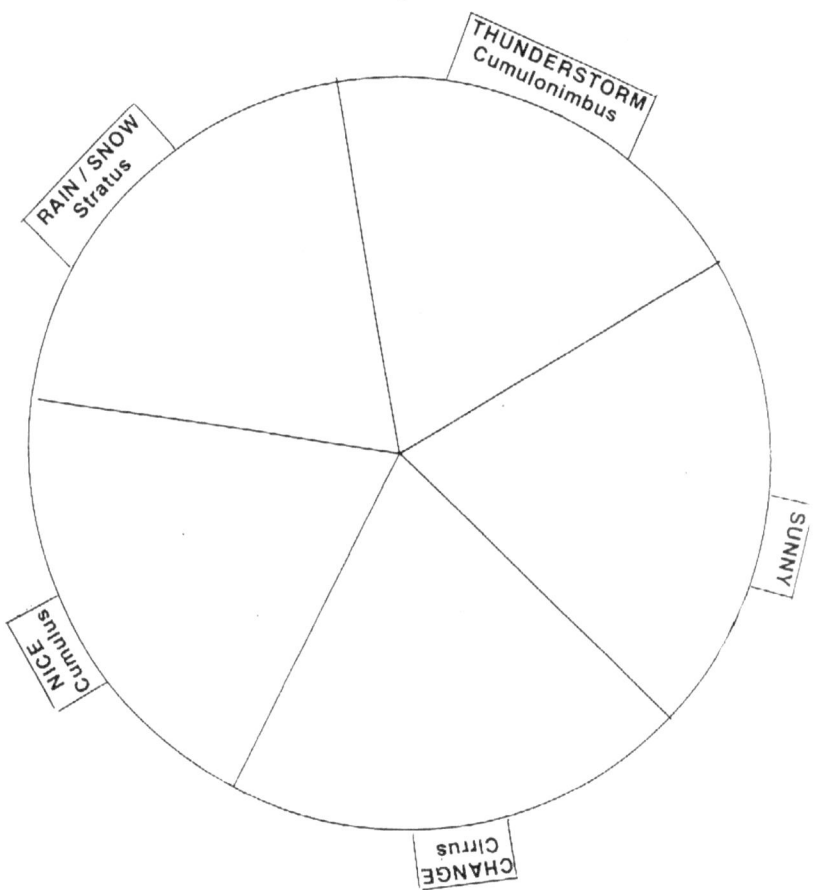

APPENDIX B

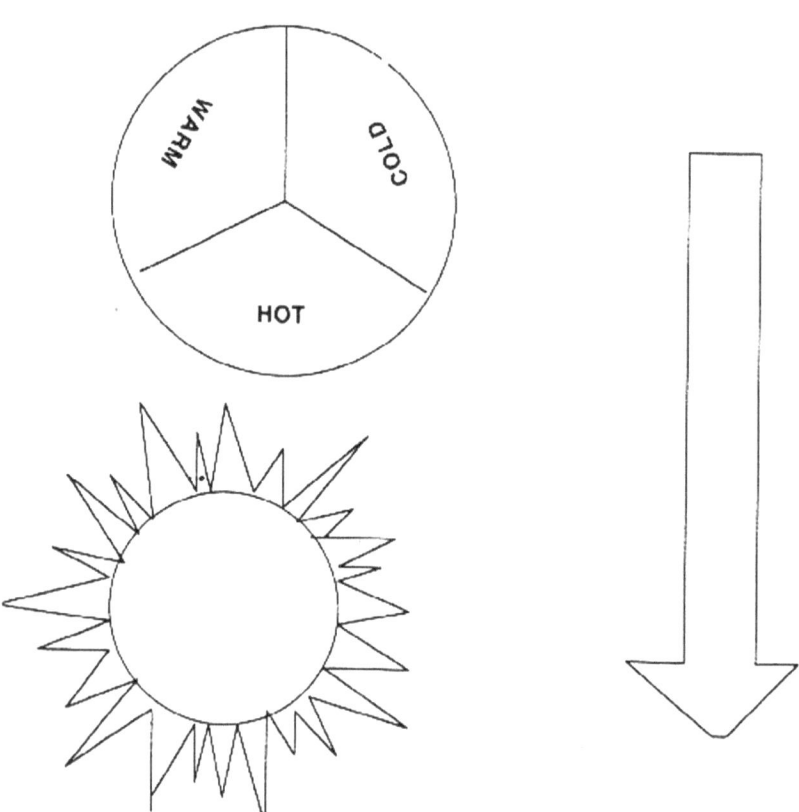

APPENDIX C

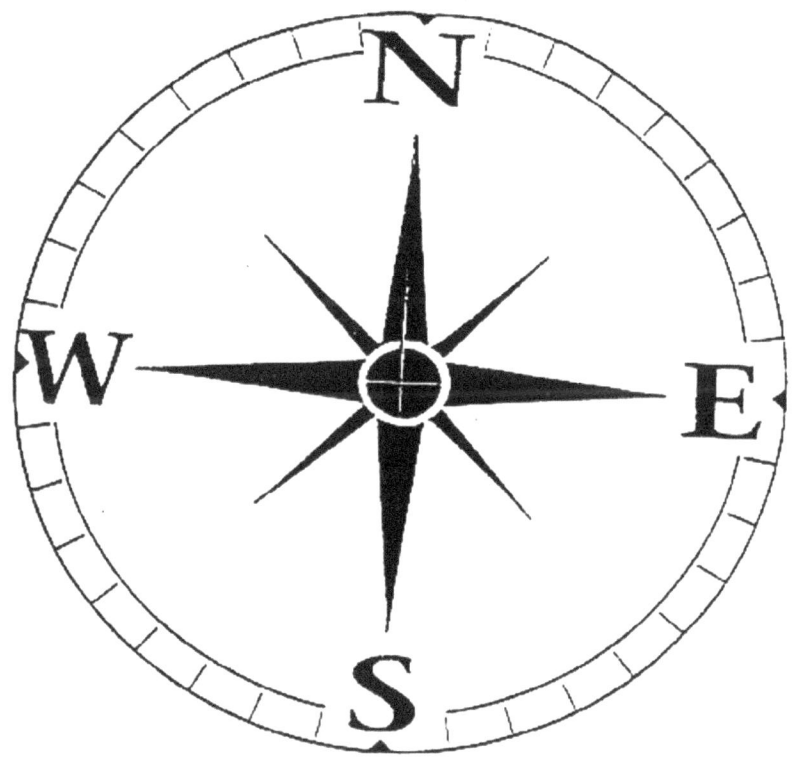

APPENDIX D

Pattern to mark holes for weathervane

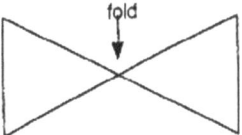

Pattern for arrow on
weathervane.

Pattern for NSEW on weathervane

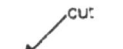

APPENDIX E

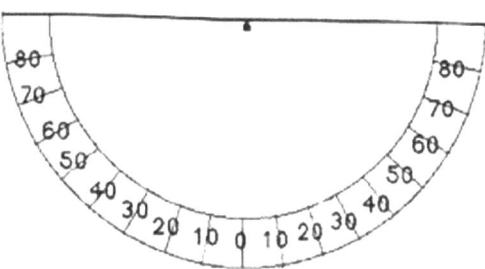

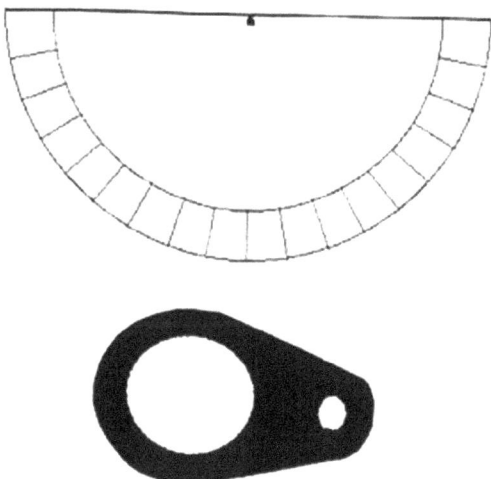

APPENDIX F

Cloud Characteristics

Cloud Name	Appearance	Associated Weather
Alto cumulus	flattened puffs	rain or thunderstorm
Alto stratus	fine, flat, gray	long periods of precipitation
Cirrus	fine, high, wispy	change
Cirro cumulus	rippled, high	nice weather, if no cahnge
Cirro stratus	continuous, thin, white	precipitation soon
Cumulus	thick, puffy	nice weather, if no cahnge in shape to Cumulo nimbus
Cumulo nimbus	large, vertical, dark	thunderstorms, heavey rain, hail
Nimbo stratus	thick, gray, covers sky	long periods of heavey precipitation
Strato cumulus	layers of mixed colors	colder, slight chance of precipitation
Stratus	low, gray, flat	light precipitation

APPENDIX G

RELATIVE HUMIDITY (%)

Dry Bulb Difference Between Wet and Dry Bulb Reading in CelsiDegrees

C	1	2	3	4	5	6	7	8	9	10
10	88	77	66	55	44	34	24	15	6	
11	89	78	67	56	46	36	27	18	9	
12	89	78	68	58	48	39	29	21	12	
13	89	79	69	59	50	41	32	22	15	
14	90	79	70	60	51	42	34	25	18	7
15	90	80	71	61	53	44	36	27	20	10
16	90	81	71	63	54	46	38	30	23	13
17	90	81	72	64	55	47	40	32	25	15
18	91	82	73	65	57	49	41	34	27	18
19	91	82	74	65	58	50	43	36	29	20
20	91	83	74	67	59	53	46	39	32	22
21	91	83	75	67	60	53	46	39	32	26
22	92	83	76	68	61	54	47	40	34	28
23	92	84	76	69	62	55	48	42	36	30
24	92	84	77	69	62	56	49	43	37	31
25	92	84	77	70	63	57	50	44	39	33

Dry Bulb Difference Between Wet and Dry Bulb Readings in Fahrenheit Degrees

F	1	2	3	4	5	6	7	8	9	10	11	12
66	95	90	85	80	75	71	66	61	57	53	48	44
68	95	90	85	80	76	71	67	62	58	54	50	46
70	95	90	86	81	77	72	68	64	59	55	51	48
72	95	91	86	82	77	73	69	65	61	58	54	50
74	95	91	87	82	78	74	70	66	62	59	55	51
76	96	91	87	83	79	75	71	67	63	60	56	53
78	96	91	87	83	79	75	72	68	64	61	57	54
80	96	92	88	84	80	75	72	69	65	61	58	55
82	96	92	88	84	80	76	73	69	66	62	59	56
84	96	92	88	84	81	77	73	70	66	63	60	57
86	96	92	88	85	81	77	74	70	67	64	61	57
88	96	92	89	85	81	78	74	71	68	65	61	58

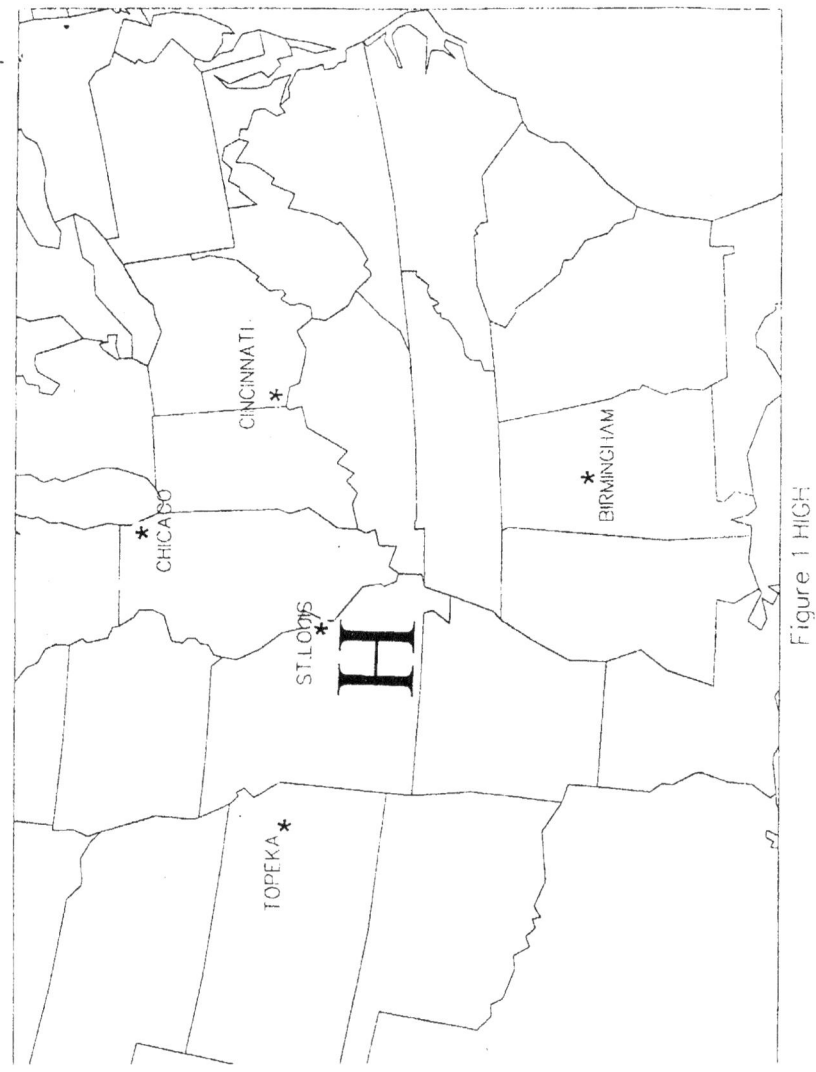

Figure 1 HIGH

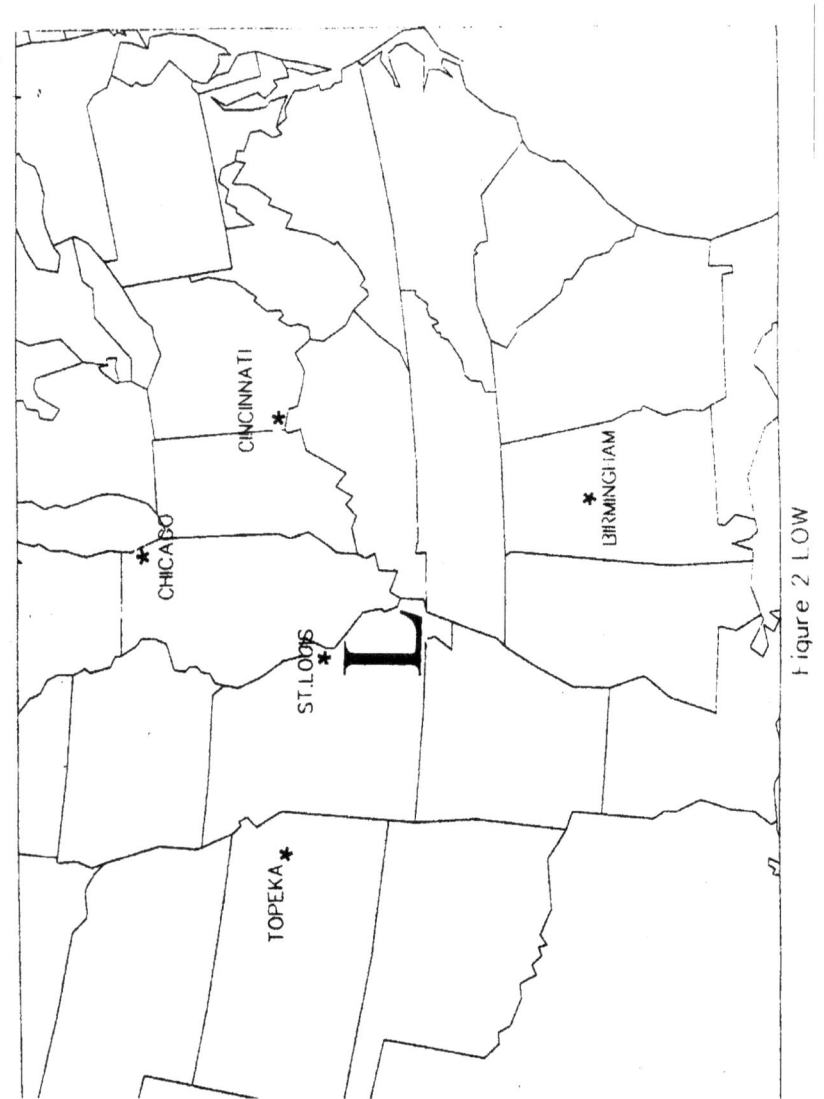

Figure 2 LOW

www.ingramcontent.com/pod-product-compliance
Lightning Source LLC
Chambersburg PA
CBHW041102180526
45172CB00001B/72

* 9 7 8 1 1 0 5 7 8 5 1 1 5 *